Let's Learn About Space

ASTEROIDS

Rebecca Kraft Rector

Enslow Publishing
101 W. 23rd Street
Suite 240
New York, NY 10011
USA
enslow.com

Words to Know

atmosphere Air that surrounds a planet.

crater A hole made by the crash of a large object, like an asteroid.

extinct No longer living.

gravity The force that attracts two objects to each other.

meteorite A rock from space that hits Earth.

orbit The path of an object as it circles another object in space.

rare Unusual.

satellite A man-made object that orbits an object in space.

CONTENTS

In space, there are millions of asteroids like this one.

What Is an Asteroid?

An asteroid is a chunk of rock. Asteroids are in outer space. They orbit the sun. They are smaller than planets. Sometimes asteroids are called minor planets.

FAST FACT

The word "asteroid" means "like a star."

These pieces of rock became asteroids when the planets formed.

How Is an Asteroid Made?

Long ago, the planets formed. There were pieces left over. These pieces are asteroids. There are millions of asteroids. All the asteroids together are smaller than the moon.

FAST FACT

Asteroids are made of stone and metal.

This asteroid is named Ida.
The small object next to it is
its tiny moon, named Dactyl.

Big, Small, and In-Between

Asteroids are not all the same size. Some are as small as a pebble. Some are hundreds of miles across.

Fast Fact

Some asteroids have tiny moons.

Some asteroids have unique shapes. This one looks a bit like a shoe.

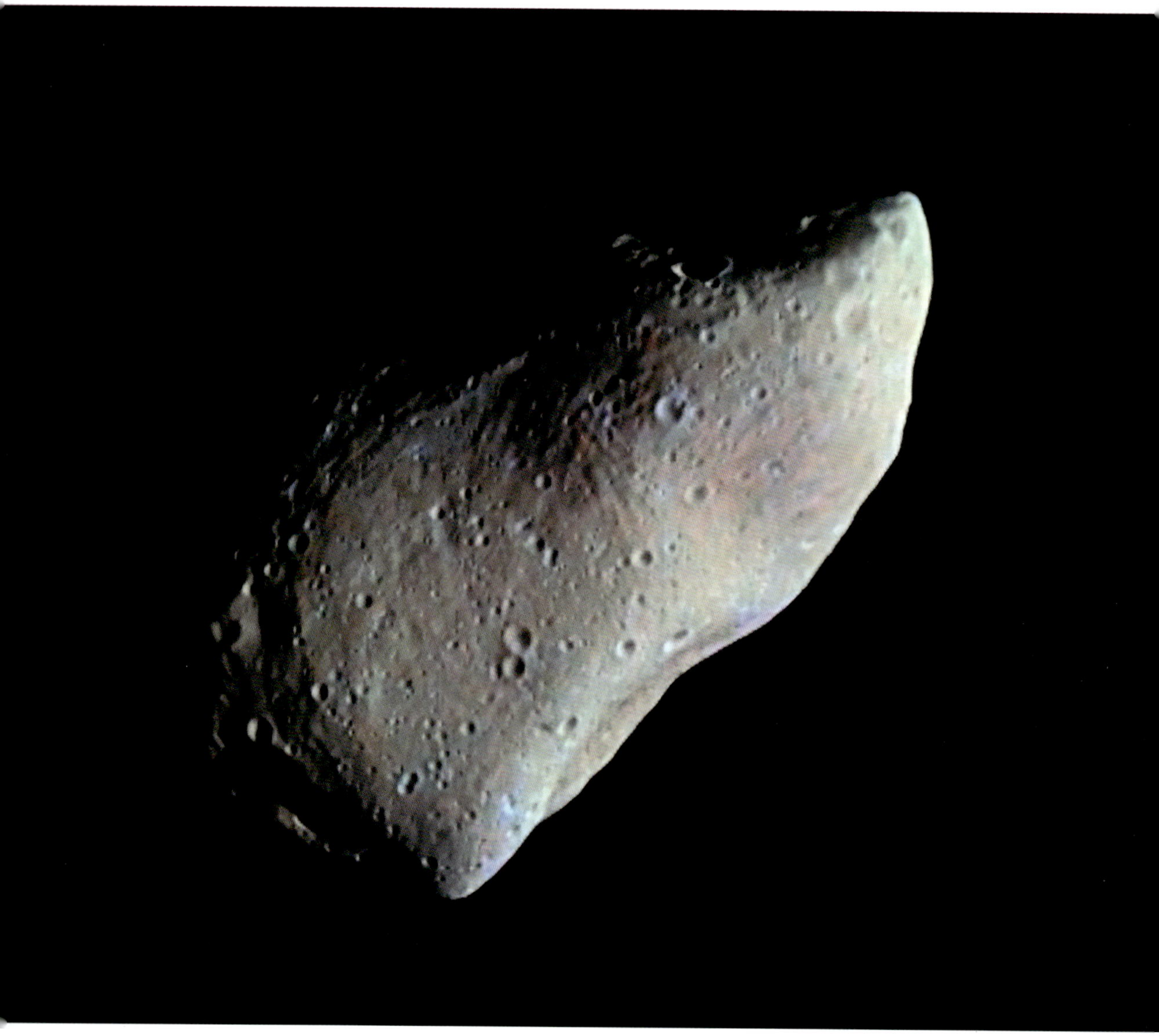

Shaping Up

Asteroids are different shapes. Some are round. Some are long. Some look like potatoes or dog bones. Larger asteroids have more **gravity**. Gravity pulls them into a rounder shape.

FAST FACT

Asteroids often have **craters** in them.

The asteroid belt is the smaller circle of rocks you see in this picture.

A Rocky Doughnut

Most asteroids are found in the asteroid belt. It is between the orbits of Mars and Jupiter. The largest object in the belt is Ceres. It is a dwarf planet.

FAST FACT

The asteroid belt is shaped like a doughnut.

These Trojan asteroids orbit with the planet Jupiter.

Sharing a Path

Trojan asteroids are not in the asteroid belt. They share the orbit of a planet. Some stay in front of the planet. Some stay behind. Earth has one Trojan.

FAST FACT

Jupiter has the most Trojans.

This meteorite was found in Russia.

Too Close?

Near-Earth asteroids pass close to Earth's orbit. Small asteroids burn up in Earth's atmosphere. Some pieces might make it to the ground. They are called meteorites.

These two lakes in Canada were formed by an asteroid hitting Earth.

Goodbye, Dinosaurs

Large asteroids can hit Earth. But this is rare. Thousands of years pass between hits. Long ago, an asteroid crashed into Earth. Scientists think this is why dinosaurs are extinct.

FAST FACT

Asteroids make craters when they crash.

This drawing shows the NEAR spacecraft next to an asteroid.

Defense!

Scientists keep track of near-Earth asteroids. They watch from all over the world. Satellites and spacecraft also help study them. The NEAR spacecraft was the first to orbit an asteroid.

Activity

Making Craters

Asteroids sometimes crash into other objects. They can leave a crater. What happens if the asteroid is big? What happens if the asteroid is small?

Let's get started!

Materials

- Shallow pan
- Flour
- Different size balls

Procedure:

Step 1: Fill the pan with flour. Pat it down.

Step 2: Drop a ball into the pan.

Step 3: Look at the crater. Is it big? Is it deep?

Step 4: Repeat with a different ball.

Do the balls make the same size crater?

You can make a "crater" with flour.

LEARN MORE

Books

Hansen, Grace. *Asteroids & Meteoroids*. Minneapolis, MN: Abdo Kids, 2017.

Riggs, Kate. *Asteroids*. Mankato, MN: Creative Education, 2015.

Yasuda, Anita. *Explore Comets and Asteroids*. White River Junction, VT: Nomad Press, 2017.

Websites

European Space Agency—Asteroids, the Minor Planets
www.esa.int/esaKIDSen/SEMCM9WJD1E_OurUniverse_0.html
Discover asteroid facts, experiments, and games.

Kids Astronomy
kidsastronomy.com/solar-system/asteroid-facts/
Explore fun facts, videos, and games about asteroids.

INDEX

Published in 2020 by Enslow Publishing, LLC.
101 W. 23rd Street, Suite 240, New York, NY 10011

Library of Congress Cataloging-in-Publication Data

Names: Rector, Rebecca Kraft, author.
Title: Asteroids / Rebecca Kraft Rector.
Description: New York : Enslow Publishing, 2020. | Series: Let's learn about space | Audience: Grade K to 4. | Includes bibliographical references and index.
Identifiers: LCCN 2018041860| ISBN 9781978507265 (library bound) | ISBN 9781978509191 (pbk.) | ISBN 9781978509207 (6 pack)
Subjects: LCSH: Asteroids—Juvenile literature.
Classification: LCC QB377 .R43 2019 | DDC 523.44—dc23
LC record available at https://lccn.loc.gov/2018041860

Printed in the United States of America

To Our Readers: We have done our best to make sure all website addresses in this book were active and appropriate when we went to press. However, the author and the publisher have no control over and assume no liability for the material available on those websites or on any websites they may link to. Any comments or suggestions can be sent by e-mail to customerservice@enslow.com.

Photos Credits: Cover, p. 1 Oliver Denker/Shutterstock.com; pp. 4, 8, 14 NASA/JPL-Caltech; p. 6 Jurik Peter/ Shutterstock.com; p. 10 NASA/JPL/USGS; p. 12 Siberian Art/Shutterstock.com; p. 16 Alexey Layeroff/Shutterstock.com; p. 18 Time Life Pictures/ The LIFE Picture Collection/Getty Images; p. 20 Hulton Archive/ Getty Images; p. 23 Lapina Maria/Shutterstock.com; interior design elements (planets) Vectomart/Shutterstock.com, (sun) Kirill Kirsanov/Shutterstock.com.